Using the Microscope

Jean Imrie MSc

A & C Black London

Acknowledgements

The author and publishers are grateful to the following for permission to reproduce photographs:

E J Arnold & Sons Ltd, fig 2; Bauch & Lomb Ltd, figs 13, 18, 19; Griffin & George Ltd, figs 1, 3, 4, 5, 17a; Philip Harris Biological Ltd, figs 11, 12, 14, 17b, 26, 28, 29, 30, 32, 33; Imperial Chemical Industries Ltd, Pharmaceuticals Division, fig 25; Opex Ltd, fig 10; Open University, figs 22, 23, 24; Osmiroid Pens Ltd, fig 7; Rosalind Price, fig 6; W R Prior & Co Ltd, fig 21; Pyser Ltd, figs 15, 20; By courtesy of the trustees of the Wellcome Institute, figs 8, 9; ©Douglas P Wilson, figs 16, 34.

Other drawings by Technical Art Services Ltd.

Published by A & C Black Ltd, 35 Bedford Row, London WC1R 4JH

ISBN 0 7136 1630 X

First published 1976

Printed in Great Britain by The Garden City Press Ltd, Letchworth, Herts.

Contents

1 The lens

It is astonishing how many people 'look' but do not 'see'. Few people look as thoroughly or as completely at objects as they might, for a casual glance is never sufficient to see all that is there. The object of attention is only really known as well as seen if time is taken to look carefully, to look from all sides, and to look often.

Normal eyesight is adequate to see a host of natural objects of astonishing beauty and complexity, yet for many years man has tried to see objects which are too small to be seen without assistance. These range from minute single cells many of which are now known to cause disease, to the complex internal structure of plants and animals. A number of instruments have been devised to examine such a variety of material. Some such as the lens are simple, whereas others such as the electron microscope are complex and costly.

From an early age we are familiar with 'glasses' or 'spectacles' which some people wear all the time and others for reading only. Glasses are the devices most frequently used for clarifying distant objects or for enlarging print in order to make reading easier. They consist of two lenses supported in a frame and worn in front of the eyes. As the eyes themselves change through age or some other cause, the lenses may have to be changed too. But even those who do not wear spectacles regularly find a single lens held in the hand or supported in some other way a useful tool in enlarging small objects and revealing unsuspected beauty in them. Such a lens or magnifier consists of a piece of glass or perspex curved on both sides (*biconvex*) and usually mounted in metal or plastic. Lenses vary in size from approximately 11 mm in diameter to 100 mm. The larger the lens the easier it is to use, and quite young children enjoy a tripod magnifier.

It may take some practice to see clearly with a small lens (fig. 1). The object to be studied must be held securely in one hand and the lens held close to the eye with the other. The object is brought up slowly towards the lens until a clear enlarged image is obtained. At this point the object is said to be in focus. Using a single lens it is obvious that only one eye at a time can see the magnified image, but if possible, both eyes should remain open.

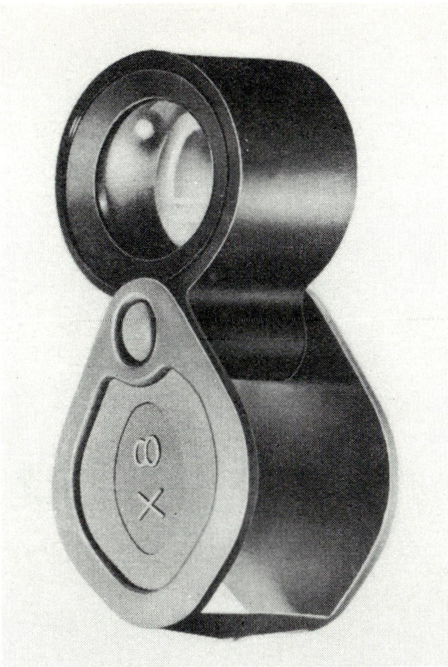

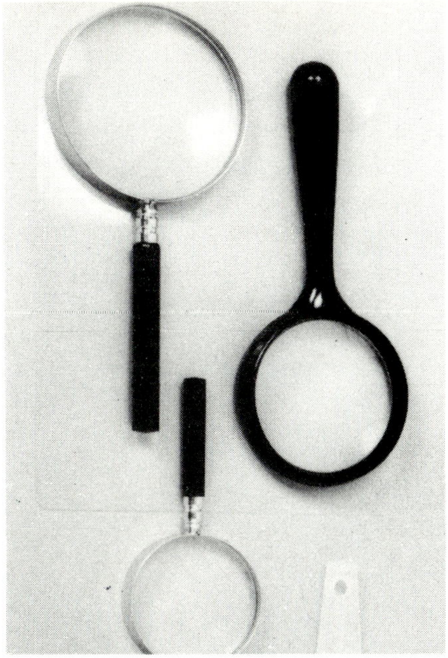

fig. 1 (left) A simple lens

fig. 2 (right) Reading glasses

With a hand-held reading glass (fig. 2) the book may be placed on the table and, with the glass in position, both eyes can be used. In the tripod magnifier (fig. 3) the lens is large and the distance between the object and the lens is fixed by the height of the legs. An object placed beneath the centre of the lens is immediately in focus and a clear image is obtained. It can be examined at leisure using both eyes.

fig. 3 A tripod magnifier

The degree of magnification varies with the size and curvature of the lens; in the more expensive ones it is usually engraved on the case. The ×8 on the lens illustrated in fig. 1 means that the linear dimensions of the object, when in focus, will all be multiplied eight times; that is, the image will be eight times longer and eight times wider than the object: its area will of course be 64 times as great.

fig. 4 A watchmaker's eyeglass

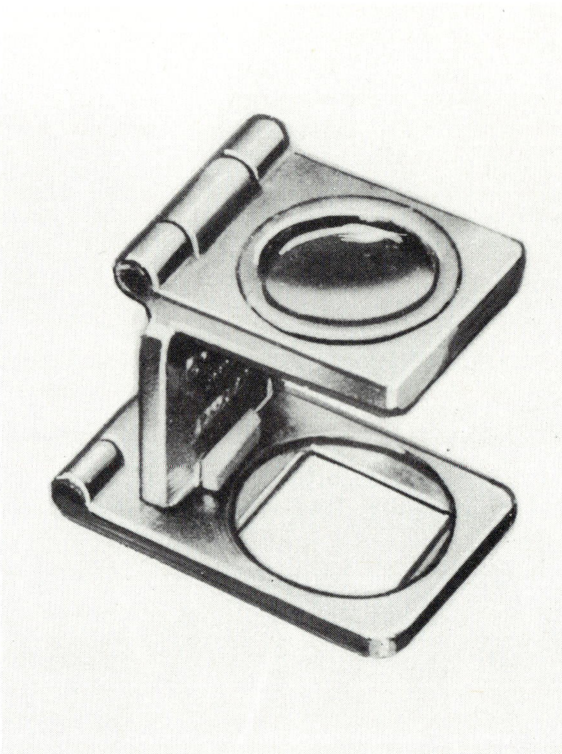

fig. 5 A piece glass, used in textile mills

Some lenses are used for specific purposes and are adapted accordingly. The watchmaker's eyeglass (fig. 4) is made to fit into the eye socket leaving both hands free for work. Its magnification is approximately ×10 and enables the worker to manipulate the tiny screws, wheels and springs of the smallest watches. The piece glass is used in textile mills (fig. 5). It has a graduated square aperture below in order to count the number of threads in a defined area; the lens above the aperture makes this easier.

The first lenses were made of glass, and the moulding and polishing of glass lenses is still an important industry. The development of plastics however, has enabled cheap substitutes to be manufactured, and acrylic material of high optical grade is being used in lens production. These lenses have the advantage of being light in weight and unbreakable, but so far they have not been extensively used in the manufacture of microscopes (fig. 6).
Transparent plastic boxes of approximately 30 mm diameter with a lens as a lid make the observation of living creatures much simpler. The degree of magnification is not large but is sufficient to examine wings, legs, gills, etc. of water creatures, insects and land animals which are otherwise difficult to control for examination with a hand-held lens (fig. 7).

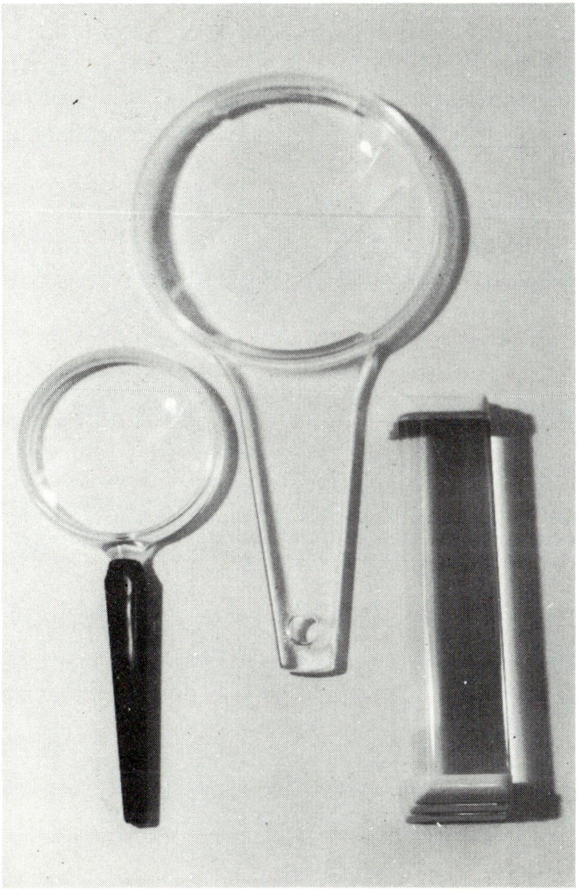

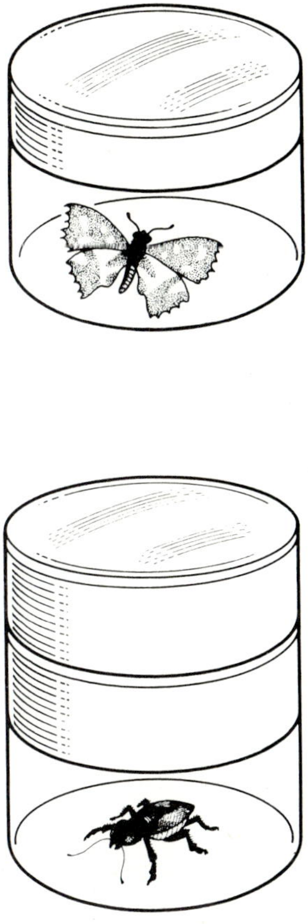

fig. 6 (left) Plastic lenses

fig. 7 (right) Plastic box magnifier

2 The monocular microscope

The single lens does not magnify to any great extent by itself and it was soon discovered that a much larger image could be obtained by using two lenses some distance apart. The arrangement of these lenses varies a little but the principle is the same in all microscopes.

The *objective lens* produces a magnified image which is further magnified by the second *eyepiece lens*. Monocular microscopes are provided with one eyepiece and the image can therefore be seen by only one eye at a time. As with the single lens, both eyes should be kept open. Binocular microscopes with two eyepieces are designed so that the image is seen with both eyes together (see page 21).

The earliest microscopes were made in Europe about the year 1600. A microscope closely resembling a modern instrument was made in England by Robert Hooke who lived from 1635 to 1703 (fig. 8). He was the first person to discover that plants were made up of millions of tiny cells, and his drawing (fig. 9) of cork cells from the bark of a tree is an historic one.

fig. 8 (left) Robert Hooke's microscope

fig. 9 (right) The first illustration of the cell structure of plants

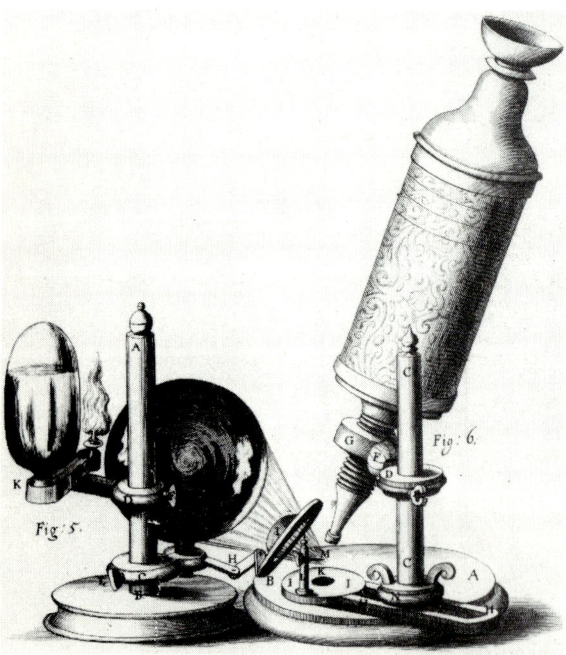

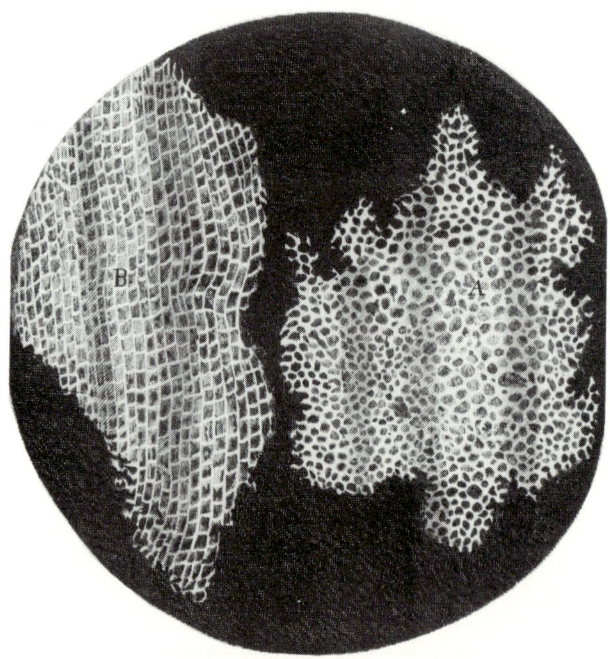

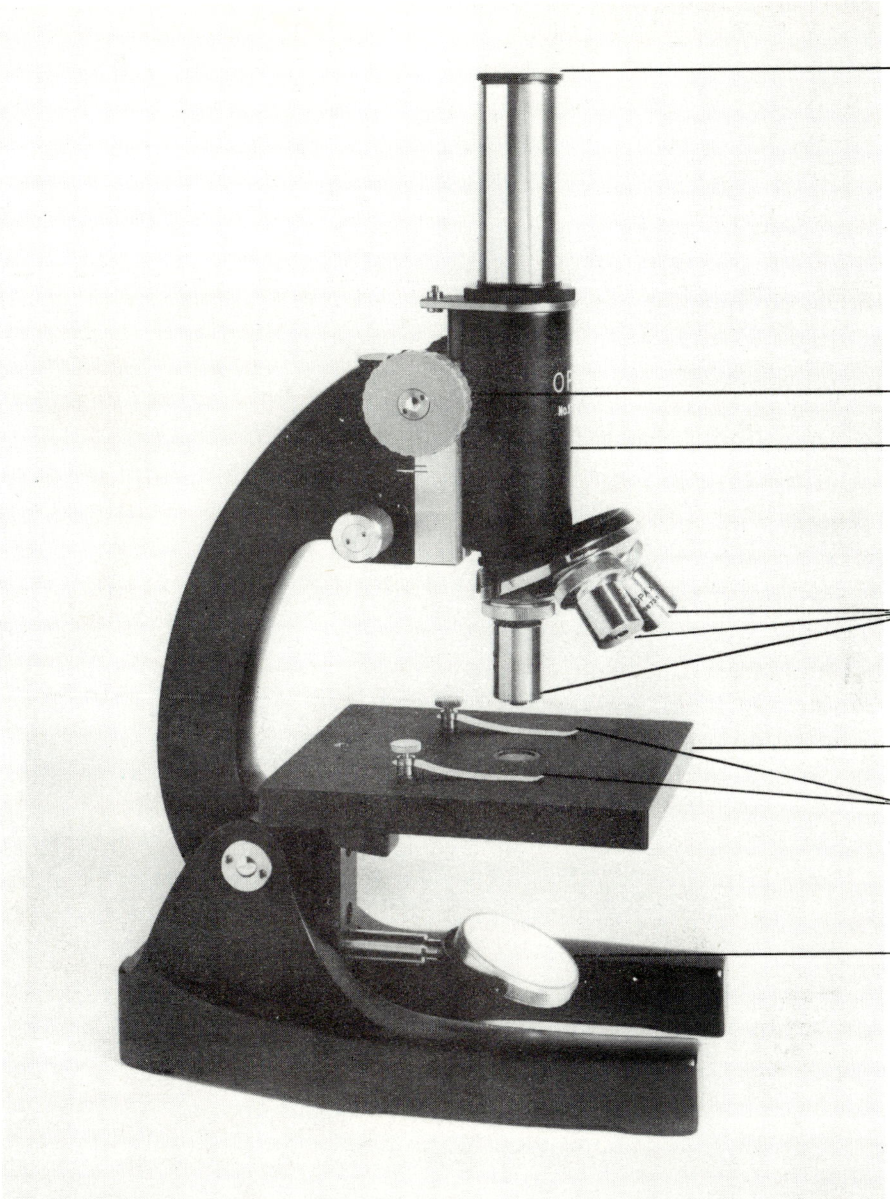

fig. 10 A monocular microscope

Figure 10 shows a modern monocular microscope. The objective lenses (three of them) are set in the objectives and these are held in the nosepiece. The eyepiece lens is clearly marked. Good light is necessary in order to see the enlarged image plainly and a mirror is placed so that it catches the light and directs it through the object, the objective lens and the eyepiece lens to the eye itself. The object must be transparent since the light must pass through it. It is held on a thin piece of glass called a microscope slide, placed on the stage of the microscope and held in place by two stage clips. The light passes through a hole in the centre of the stage.

A beginner might use a prepared slide of some familiar object for the first examination but if this is not available a few grains of salt or dry sand, a small feather, some strands or hair of wool will do. Place the object in the centre of the clean slide, position the slide on the stage and secure it by the stage clips.

To see a distinct image of a very small object, lighting and focussing are very important. In the model illustrated, focussing is done by directing the light through the object by means of the mirror, placing the shortest objective (which gives the lowest magnification) over the object and using the coarse adjustment screw. It is wise to look at the side of the microscope, turn the screw until the objective lens is close to the object and then, looking through the eyepiece, turn the screw so that it brings the microscope tube upwards until the image can be clearly seen. The object is then in focus. The contents of the slide can then be carefully examined by dislodging the stage clips and moving the slide gently over the stage. Small movements of the slide produce much enlarged movements when seen through the microscope, and the object and its image move in opposite directions. You will notice that if the slide is moved a little to the left, the image seen down the microscope moves to the right. It takes considerable practice to position the object in the centre of the field of view for further examination at a higher magnification.

In order to produce a larger image the nosepiece carrying the objectives is rotated until the middle-sized objective is over the centre of the stage. Again looking at the microscope from the side, use the coarse adjustment screw to position the objective close to the object on the slide. Looking once more down the microscope tube, bring the tube upwards until the image is clearly seen. It may now be necessary to use the fine focussing adjustment screw at this stage to obtain the most distinct image possible.

A still greater degree of magnification can be obtained by rotating the nosepiece until the third and longest objective is over the centre of the slide.

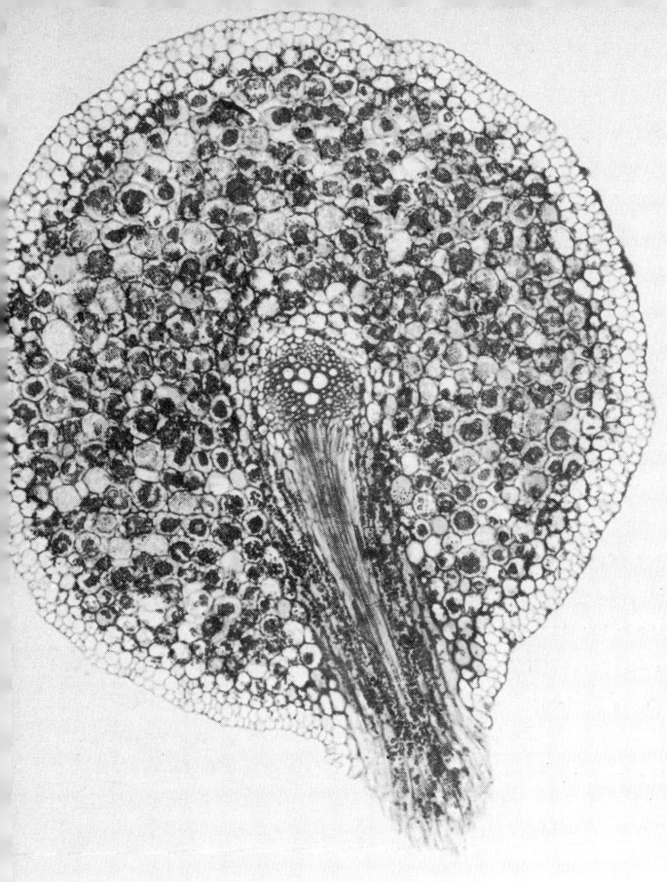

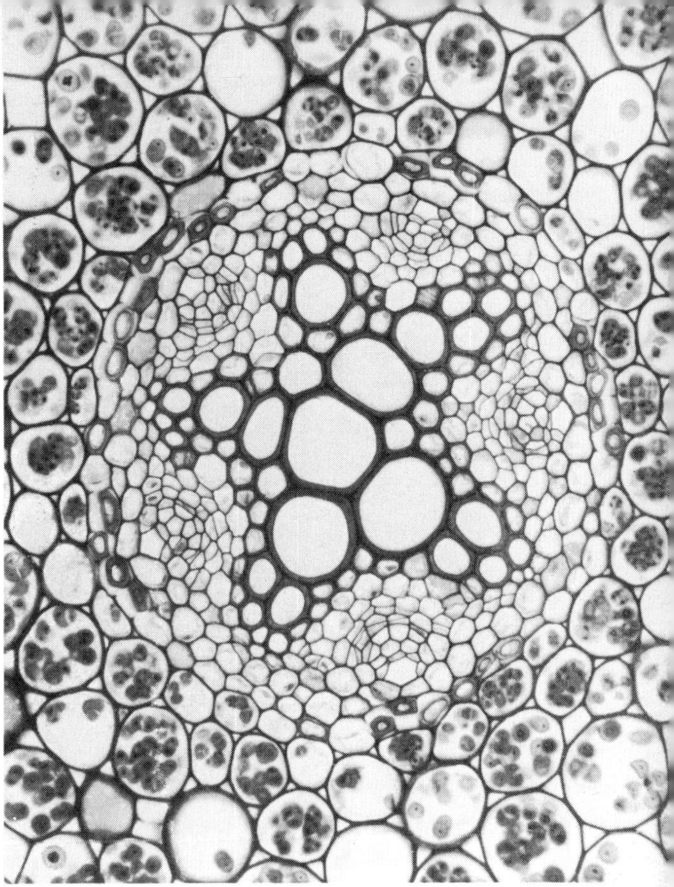

fig. 11 (left) A buttercup root, showing a lateral shoot, at a magnification of ×60

fig. 12 (right) Buttercup root at a magnification of ×300

Great care must be taken in using this objective; in order to focus the object correctly, the objective lens will be very close indeed to the slide and may accidentally be pushed into it and destroy it. The fine adjustment screw should be used for accurate focussing (figs. 11 and 12).

The degree of magnification varies with the objective used. The markings on the side of the objective will show this. Recent models indicate the actual degree of magnification, ×10, ×40 or × 100. Some older models indicate the distance the objective lens should be from the object on the slide when in focus, and the markings might be ⅔", ½" or ⅛". A comparison can be made, for the ⅔" objective gives a magnification of ×10, the ½" objective gives a magnification of ×14 and the ⅙" objective a magnification of ×40.

Whatever the degree of magnification of the objective lens, it is increased again by the degree of magnification of the eyepiece lens.

This lens may be ×2, ×4 or ×10, so the final degree of magnification is obtained by multiplying one by the other. The objective commonly used for low power work is ×10 and with a ×10 eyepiece, the object will be magnified $10 \times 10 = 100$ times. Using a ×40 objective and the same eyepiece, the degree of magnification will be $40 \times 10 = 400$ times. As with the hand lens, it is important to remember that it is the linear dimensions which are magnified. Remember also that the eyepiece lens magnifies the image and does nothing to make it clearer. The objective lens is therefore the more important one and must at all times be kept clean, dry and free from scratches.

The actual size of microscopically small objects is not easy to determine. It is measured in micrometres. Each micrometre is one-millionth of a metre (10^{-6} m) or one-thousanth of a millimetre, (10^{-3} mm) and it is denoted by the symbol μ. Some general idea of size can be obtained by placing a transparent plastic ruler beneath the low power objective of the microscope and looking through the eyepiece at the graduations, especially at the millimetre markings. This gives a rough idea of how many millimetres there are in the field of view. Further examination of the object may make it possible to estimate its size in millimetres or fractions of a millimetre. To determine the size more accurately than is possible by this method, a *graticule* inserted in the eyepiece, or a *stage micrometer* must be used. A graticule is a measuring device and consists of a piece of film on which is marked a 10 mm length divided into millimetres (10^{-3} m) each of which is divided again into 10 parts. Each of the smallest divisions therefore measures 100 micrometres. The film is carefully cut to fit into the eyepiece of the microscope and the lens is screwed back into place. The observer can see the markings as he examines the slide and can use them to give a more accurate estimate of the actual size of the structures seen.

A stage micrometer is an ordinary microscope slide with a scale etched on it. The scale is 10 mm in length with divisions which may be as small as 0·01 mm along it. The scale is placed on the microscope stage and the instrument is focussed on the scale with the high power objective. Using the 0·01 divisions to measure the field of view, the size of the object can be estimated as before.

Not every monocular microscope is exactly like the one in fig. 10; there are differences to be found in the method of focussing, in lighting and in the construction of the objectives. There are additions such as condensers and filters. The standard length of a microscope tube is 160 mm and most microscopes have a standard tube width and standard screw thread for holding the objectives in the nosepiece. It follows that eyepieces and objectives from different instruments are inter-changeable, but where possible confusion of parts should be avoided. If additions to an instrument are required they should be obtained from the maker of the original instrument.

The eyepiece

Most microscopes are supplied with one eyepiece, usually ×10, but others will be available. In addition to its main function of enlarging the image formed by the objective, the eyepiece also brings the light rays to a focus at the eye point. Ordinary eyepieces are called Huyghenian eyepieces. There are others called 'wide-field' eyepieces which extend the actual field of view visible to the observer. For example, with one instrument using an ordinary ×10 eyepiece, the field of view is 3·4 mm in diameter whereas under the same conditions but with a wide-field eyepiece the diameter is 4·5 mm. Another advantage of a wide-field eyepiece is that the focus at the eye-point is higher, and the eye need not be quite so close to the eyepiece lens itself. This helps those observers who need to wear spectacles for work of this kind.

Some eyepieces have a built-in pointer, a very tiny needle-like spine which can be rotated to draw attention to aspects of the object being observed. Without it, it is sometimes difficult to point out particular areas of the object to anybody else who may be examining the slide.

Zoom lens eyepieces are now becoming available. They can be supplied separately for some instruments, but in others the zoom lens eyepiece is an integral part of the instrument. Fig. 13 shows a 'zoomscope'. The slide is placed on the microscope stage as in all other microscopes, the mirror is adjusted to give maximum illumination, and the knob at the side is turned to bring the object into focus. The model shown has a minimum magnification of ×25. As the object is being observed through the eyepiece, the eyepiece itself is rotated and the degree of magnification is increased until a maximum of ×100 is obtained. There is some reduction in the field

fig. 13 Zoomscope

of view and some loss of focus which is easily corrected by using the focussing knob. Since the nosepiece is not rotated to change the objective lens, the image is never out of sight. This instrument is especially useful for those who find it difficult to appreciate degrees of magnification. A similar instrument provides magnification continuously from ×50 to ×200.

The objective

The *degree of resolution* of the whole instrument depends upon the quality of the objective lens. The resolution is the clarity of the image under high magnification when the object is correctly illuminated. To obtain the best image, the objective lens must admit a converging beam of light. In most instruments this beam is provided by means of the sub-stage mirror. The maximum width of the cone of light which the lens will admit is called its *numerical aperture* or NA.

The numerical aperture is often marked on the objective itself, for example, $\times$10/0·25 or $\times$40/0·65. The bigger the NA, the greater is the capacity of the objective lens to show up minute detail in the object being examined. Fig. 14 shows a diatom, *Navicula lyra* photographed with three objectives in turn to show the effect of increased numerical aperture on resolution.

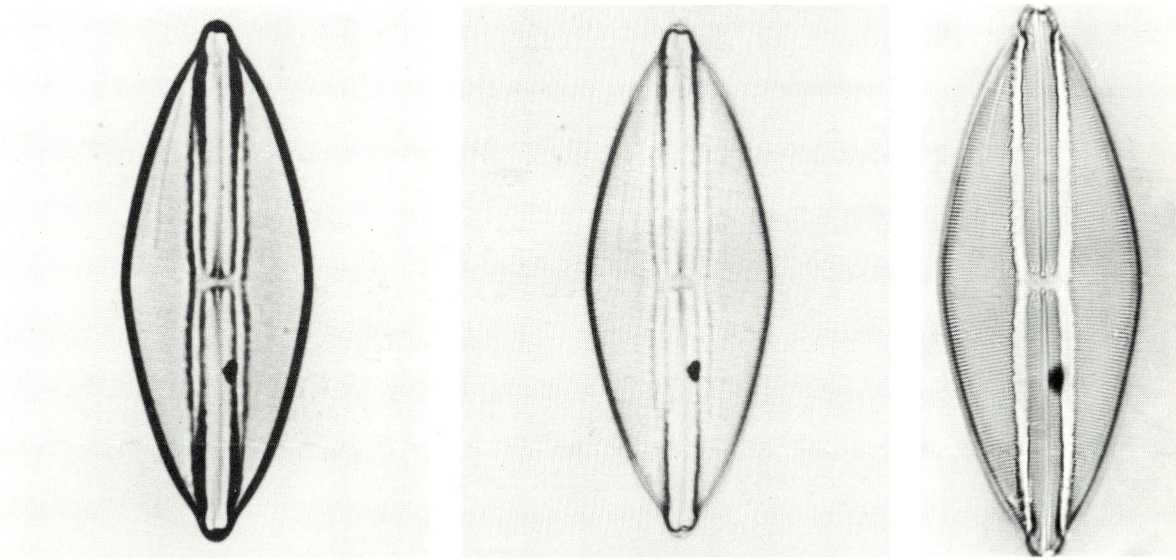

fig. 14 *Navicula lyra* photographed with three objectives. From left to right these are NA 0.25, NA 0.45 and NA 0.65

For the examination of specimens with objectives up to $\times$20, a concave mirror beneath the stage is sufficient to produce the converging beam of light. At greater magnifications a sub-stage condenser is needed (see page 16).

In general, objectives with magnifications of $\times$10, $\times$20 and $\times$40 will be adequate for most purposes, but for the examination of minute organisms such as bacteria or to see the details of nuclear divison, an objective of $\times$100 will be required. For a number of reasons such an objective lens has to be used with oil instead of air between it and the top of the cover glass which protects the specimen. Before using an oil immersion objective, the area of the slide for detailed study must be selected by preliminary examination under the high power objective. Then a tiny drop of immersion oil (usually cedar wood oil supplied with the objective lens) is placed on the slide and the oil immersion objective is lowered into position. As before, it is essential to watch from the side until the objective lens touches the oil and almost touches the cover glass. Then, looking through the microscope, the fine adjustment screw is used to move the tube slowly upwards until the object is as clearly visible as possible. Using a ×10 eyepiece, the object will be seen at a magnification of ×1000. The objective lens must be cleaned after use with a soft cloth to prevent the oil from sticking to it and if necessary it may be further cleaned with a little ether or xylol.

With high power objectives such as ×40 and ×100, the distance between the lens and the cover glass is so very small that there is some danger of breaking the cover glass and at the same time scratching the lens itself. To prevent this, these lenses are sometimes mounted in spring-loaded objectives. Alternatively, there may be a stopping device on the microscope tube which makes it impossible for any objective lens to reach the cover glass.

The design of objectives remained unchanged for many years, but recently at least three manufacturers of microscopes have produced new patterns (fig. 15). In general these simplify the operation of the microscope and make it less liable to damage. The objectives are not all removable, and with a ×10 eyepiece the instrument illustrated provides magnifications of ×40, ×100 and ×200 only.

The condenser

A sub-stage condenser is needed for work at high magnification. This is a lens system which focusses light on to the object and enables detail to be seen more clearly. In other words, it improves the degree of resolution. Some microscopes have a simple converging lens inserted into the hole in the centre of the microscope stage. The sub-stage mirror must be turned so that its plane surface directs the light through this simple condenser. The beam is then focussed by the condenser lens through the object on the slide and into the objective lens. More complex instruments have a condenser system consisting of two lenses. This system is known as an Abbé condenser, after its originator. Such a lens system has to be exactly focussed and any instructions supplied with the microscope should be carefully followed. As a general rule, the top lens of the Abbé condenser should be just below the level of the stage where the slide will rest. As before, the plane side of the sub-stage mirror is used to reflect the light through the system.

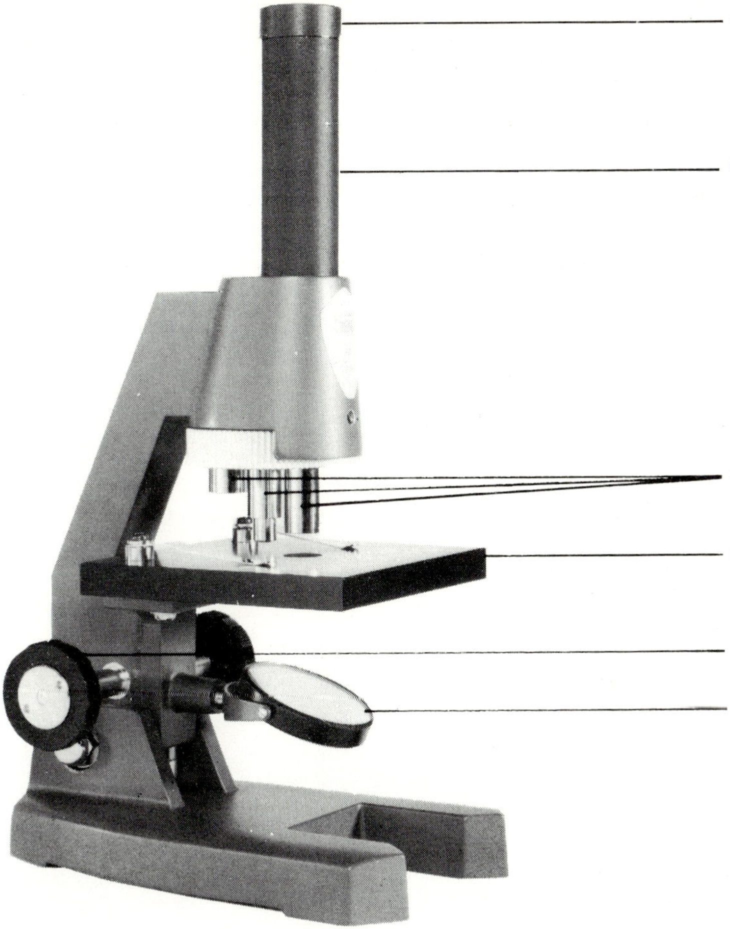

fig. 15 A microscope with the new style of objectives

Focussing

The greatest care must be taken in focussing the object being examined because only then is the object seen with the greatest possible clarity. In most instruments this is done as already described, by raising and lowering the microscope tube until the objective and eyepiece operate together. Some microscopes designed recently have a different method of operation: the objectives and microscope tube are fixed and the whole microscope stage carrying the slide to be examined moves up and down by means of a screw at the side (fig. 15). By providing a 'stop', the manufacturers have tried to ensure that the stage cannot push the slide on to the high power objective.

Lighting

It is often convenient to use the microscope in daylight; ordinary window light is adequate for this but bright sunlight should be avoided. The light is 'caught' on the mirror shown in figs. 10, 13, and 15, and is reflected upwards through the stage, through the object on the slide, through the objective and the eyepiece lenses, to the eye. The mirror is plane on one side, concave on the other, and it can be rotated in its gimbals to obtain the maximum amount of light. The concave surface focusses light on to the object and is adequate for low power work. The flat or plane surface directs the light rays into the focussing condenser when higher magnification is required.

Sometimes the light is too intense for comfortable working and many microscopes are fitted with a sub-stage diaphragm which can be used to control the width of the cone of light. The simplest form of diaphragm consists of a black disc in which there are five holes of different diameters. This disc is easily rotated horizontally to control the amount of light entering the lenses. It is a good idea to let in the maximum amount of light to begin with and then progressively to cut down. A more complicated iris diaphragm is fitted to more expensive models but its purpose is the same.

fig. 16 Living marine plankton, seen at approximately ×15, using a dark ground illumination

Some objects, such as tiny pond creatures and some animal cells, are so transparent that they can barely be seen under the microscope even when the light is adjusted to a minimum. The use of dark ground illumination helps to make them more clearly visible (fig. 16). The mirror or condenser is adjusted so that the widest possible cone of light is focussed through the slide, and then a small opaque object, for example a piece of card, is held between the mirror and the stage so that the centre of the light cone is prevented from passing into the objective lens. The transparent objects, however, will scatter light from the outer regions of the light cone and will show up white against a dark background. It is worth experimenting to see what effect this has though some experience is required to exploit this method to the full.

Another method of observing living cells which are always transparent, is by an arrangement known as phase contrast. This requires a sub-stage condenser fitted with a phase annulus and objectives fitted with phase plates. Setting up such an instrument requires care and experience and is beyond the scope of this book. It is sufficient to say that by this means, nuclei in a state of active division can be seen. Experienced observers may wish to explore this method of using the microscope. Detailed instructions for using phase contrast are available from the manufacturers of the accessories.

Although daylight is often preferable, a good light from the sky is not always available. The sub-stage mirror ensures that artificial light can be used and obviously an electric lamp of some kind is the most convenient source. There are a number of lamps to be found but the best is one which gives a good beam of light on to the sub-stage mirror but which does not dazzle the observer. The even illumination from an opal bulb is preferable to any other. Care must be taken to see that the voltage of the lamp and the voltage of the supply available are appropriate. Some laboratories are provided with transformers so that a 12 volt or 24 volt lamp is all that is required. In other cases the lamp holders are provided with individual transformers and can be connected to the mains supply. Bench lamps are usually adjustable (fig. 17, page 20) and may be used for more than one purpose.

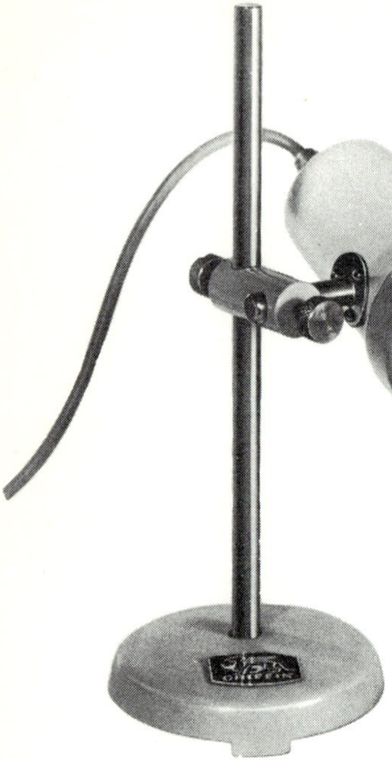

fig. 17 Two adjustable microscope lamps

So far, only sub-stage illumination has been mentioned because the microscope is most frequently used for transparent objects mounted on slides. The light is transmitted through the object to the eye. Occasionally however, small objects may be put on to the slide and the light directed on to them, that is, on to the top of the slide. The light is then reflected from the object to the eye. To be of maximum use then, the artificial light source must be capable of illuminating objects from below by transmitted light and from above by reflected light. The small bench lamp illustrated in fig. 18 is not adjustable in the same way as the others but it can be used in two positions as shown.

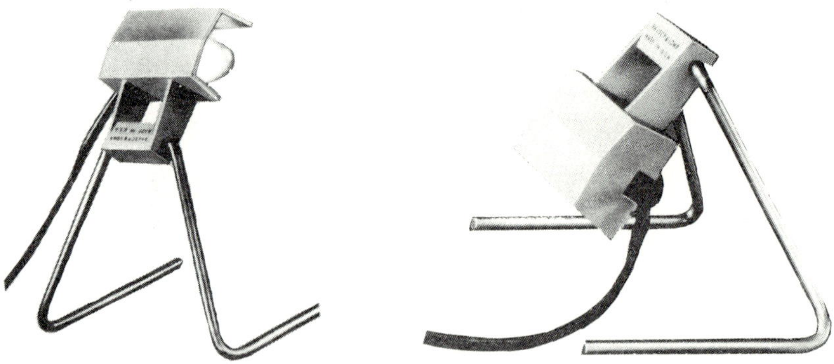

fig. 18 A 'Rite-lite' illuminator, in the positions for reflected light (left) and for transmitted light (right)

Some microscopes have the light source built into them. For some people this is a convenience but there are certain drawbacks. After prolonged use, the stage becomes uncomfortably hot and causes water-mounted specimens to dry out. The instrument cannot make use of natural light and if it is to be used for specimens requiring reflected light, an additional lamp is needed and the integral lamp switched off.

3 Some other microscopes

The binocular microscope

The binocular microscope is a never-ending source of delight. Its great advantage over the monocular microscope is that objects are seen by reflected light, and because both eyes are used they are seen in three dimensions. Details of the compound eyes of insects, the mouth parts of spiders, wings and legs of bees, flies, moths and butterflies are wonderfully revealed. Complete plants of lichens and mosses, the tiny florets in the centre of the daisy, rocks and crystals, can all be seen in their full colour and complexity. Good instruments are expensive but a great deal of pleasure can be had from the relatively cheap model illustrated in fig. 19. Magnification is fixed at ×15 but this is sufficient for many purposes. The two eyepieces can be adjusted for varying eye widths and the focussing knob moves the stage plate up and down. Quite large objects can be examined with this instrument. It needs good daylight or top (reflecting) light for most effective use but the simplicity of its design and operation makes it a most appropriate instrument for young people.

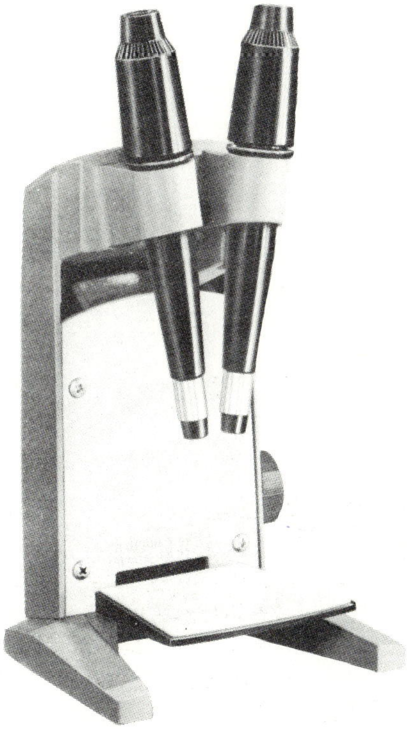

fig. 19 A binocular microscope

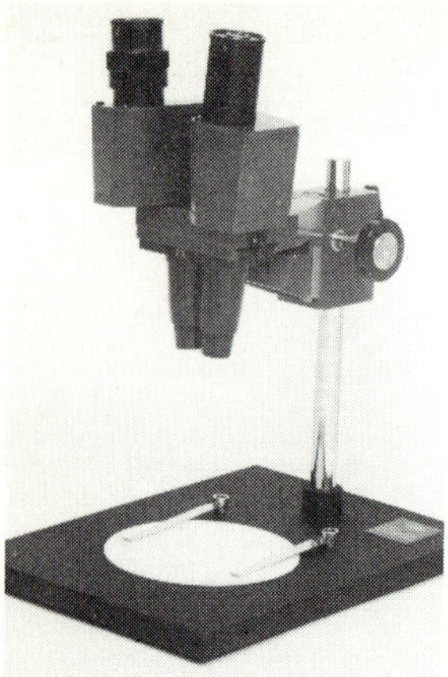

figs. 20 & 21 Low power binocular microscopes, or stereomicroscopes

Another model is shown in fig. 20. Focussing for individual eye differences is possible in this instrument. There are two paired eyepieces, ×10 and ×15 and these may be Huyghenian or wide-field. The paired objectives are ×1 and ×2 so a maximum magnification of ×30 can be obtained.

A more expensive instrument with built-in illumination and an especially heavy base is shown in fig. 21. It can be used over a dish containing pond water or leaf litter or over a dissecting dish. Quite large specimens of rock and mineral can be used under it too. Accessories provide for a maximum magnification of ×35. Focussing is by means of a knob controlling a rack and pinion, and the whole arm can be raised or lowered on the pillar. Such an instrument can reveal unsuspected beauty of pattern and shape in an infinite variety of objects and will give hours of pleasure to an observer.

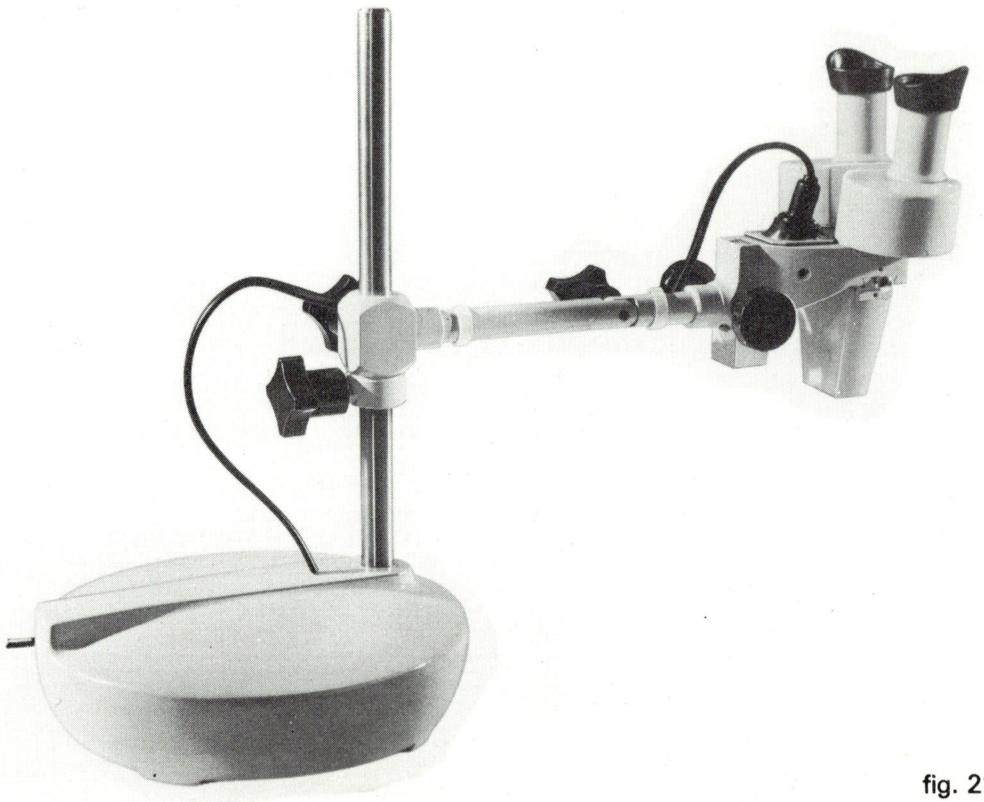

fig. 21

The Open University microscope

In a book of this nature it is impossible to give details of all the microscopes available. On the whole they all operate upon the principles described in the previous chapter and resemble each other in many ways. An important exception is the microscope designed for the students of the Open University by Dr John McArthur. This instrument looks quite different from the models illustrated so far and is shown in fig. 22. It is a rectangular solid in shape, measuring 125 mm ×75 mm ×25 mm. Because of its compactness and small size it can be carried easily in the pocket. It works well in daylight and in artificial light and it is also provided with an internal light supplied by two 1·5 volt batteries. The standard version of the instrument provides magnification of ×80 and ×200; the triple objective version gives an additional magnification of ×400. A polarising version is used for the examination of thin sections of rock (see page 38).

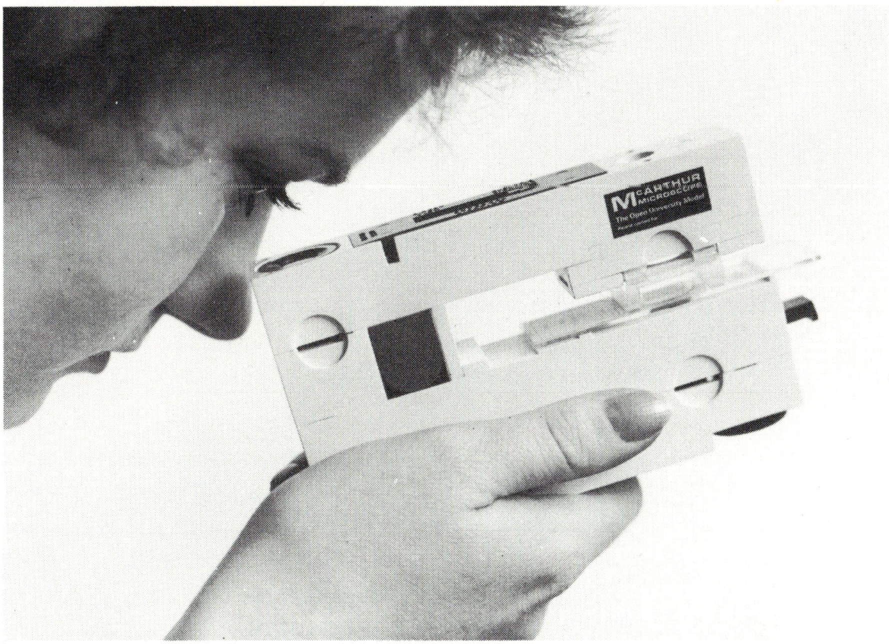

fig. 22 The Open University microscope

In the standard model, the slide to be examined is placed on the microscope stage with the cover glass facing downwards, and light passes through it as before. However, instead of a vertical course from the mirror through the objective lens, object on the slide, the microscope tube and eyepiece to the eye, the light rays are twice reflected by prisms within the instrument before passing through

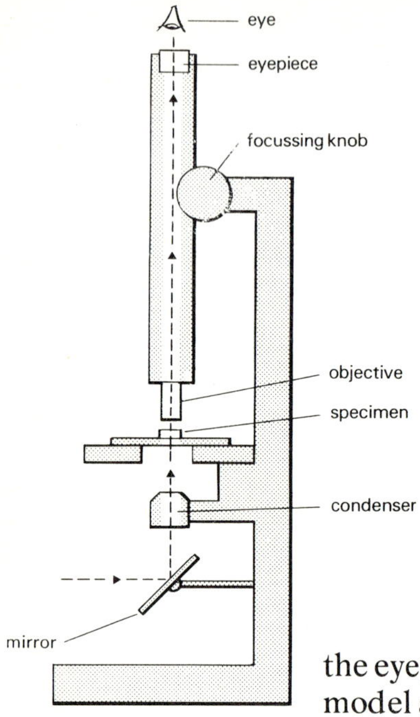

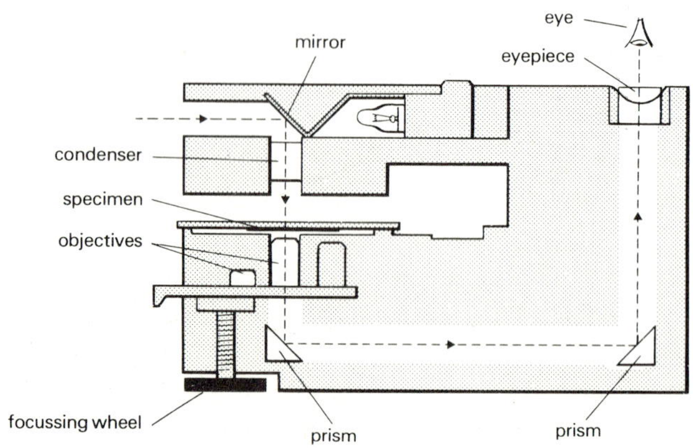

figs. 23 & 24 A comparison between a conventional microscope and the Open University microscope

the eyepiece to the eye. The paths of the rays through a conventional model and through the Open University model are shown in figs. 23 and 24. The standard model has a single eyepiece, ×10, and two objectives, ×8 and ×20. Focussing is by means of a small wheel beneath the instrument.

To someone already familiar with the relatively heavy and stable instrument, this model seems very light and it takes time to become accustomed to handling it. But its portability and ease of operation make it a most useful instrument for field work.

The electron microscope

Scientists have tried repeatedly to increase the degree of magnification of their microscopes, but a stage is reached when further magnification does not make the image any clearer. As mentioned earlier, it is the degree of resolution which is important. To see any object under the microscope, light is essential. When a light wave meets an object a deflection results and this enables us to see the cause of the deflection. But if the units making up the object are very, very small or close together, they do not individually cause any deflection of the light waves and so are not individually visible to us.

In 1935, an electron microscope was first constructed. Instead of a beam of light, an electron microscope sends a beam of electrons through the object to be examined. As the wave length of this beam is very short, visible disturbances are caused by minute objects or parts of objects hitherto indistinguishable. Whereas a magnification of ×2000 is practically the limit with an optical microscope, objects are commonly magnified 50 000 times with the electron microscope (fig. 25).

fig. 25 The electron microscope is expensive to install and complicated to operate, and is found only in universities and research laboratories

Toys

As with everything else, the better the instrument the more it costs. Because so many people, many of them children, find such pleasure in seeing objects through the microscope, a number of microscope-like toys are to be found in shops. These sometimes look like diminutive microscopes or they are described as microscopes. They are usually disappointing in performance. A young person will derive more satisfaction from a good lens than from a poor substitute for a microscope. A real microscope is more likely to be found in a shop dealing in optical instruments than in a stationers or fancy goods store. Catalogues of reliable instruments are available from educational suppliers of scientific equipment.

4 Preparing objects for the microscope

Because the monocular microscope depends upon light passing through the object, it follows that such objects must be very thin or transparent. Some objects, for example the proboscis of the blow fly (fig. 26), are transparent in parts and the microscope enables the observer to see details of hairs, bristles and veins which cannot be seen by the eye alone. Minute pond creatures which appear as moving specks in the water can be seen in great variety by simply placing a drop of pond or ditch water on a slide and focussing correctly. But the internal structure of leaves, roots or bones can only be seen by making thin preparations of the material and mounting them for observation.

fig. 26 Proboscis of a blowfly at a magnification of approximately ×40

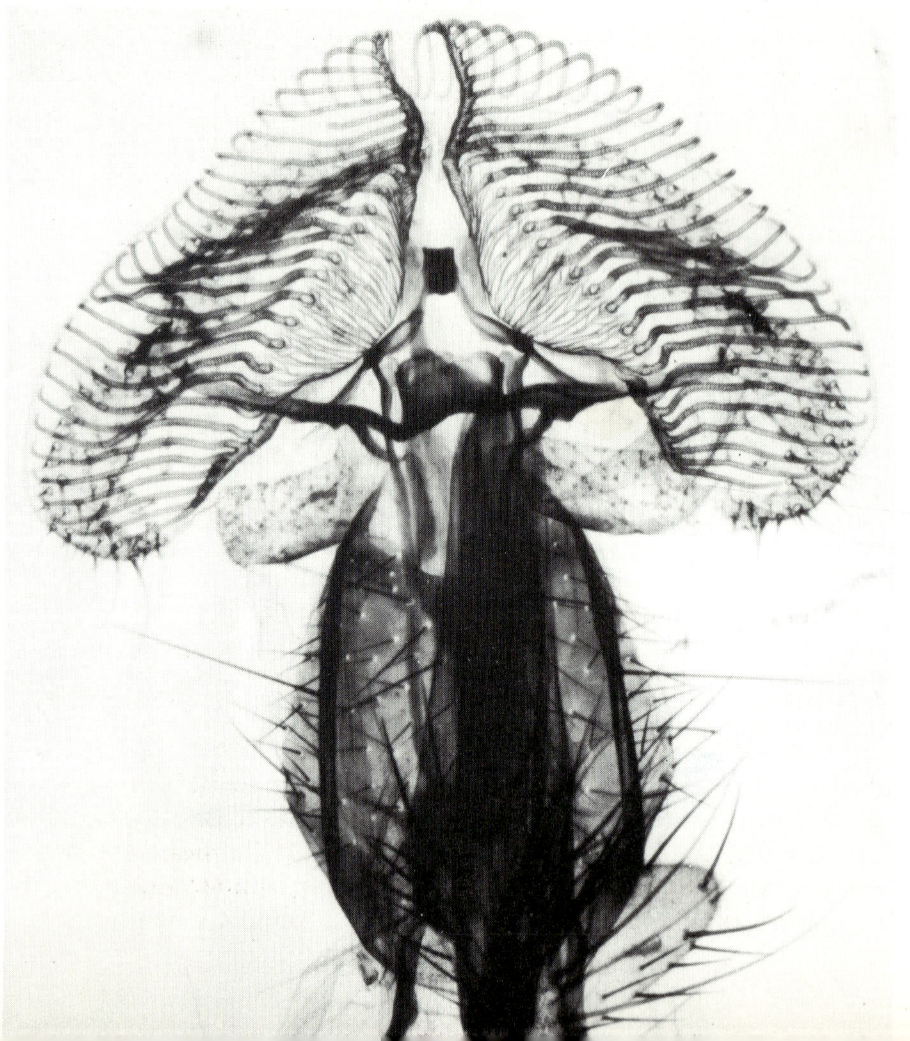

It has already been suggested that the beginner might look first at a few 'dry' objects on a slide, but the great bulk of microscope work is done with objects mounted in water or in some other fluid so that they do not dry up or shrivel. The liquid is placed centrally on the slide, the object is transferred to it and is then covered with a cover glass or cover slip to prevent the objective lens coming into contact with the liquid. The simplest way to make such a preparation is to mount a piece of the skin (epidermis) of a leaf. If you peel an onion, you can easily separate the very fine transparent skin covering the overlapping bulb scales (which are really the bases of leaves) from the fleshy material. Cut away a tiny piece of this with a sharp knife or scalpel. Using a teat pipette or a piece of glass tubing, place a single drop of water in the middle of the slide and using either fine forceps or a mounted needle, transfer the piece of epidermis into the drop of water. An iris leaf is equally suitable: to peel off a piece of the epidermis, snap the leaf by bending it over a finger and tear off a piece of the transparent epidermis with a pair of forceps. If the leaf does not break easily, a cut can be made in the leaf with a sharp knife and a piece of the epidermis gently pulled away. Cut off a

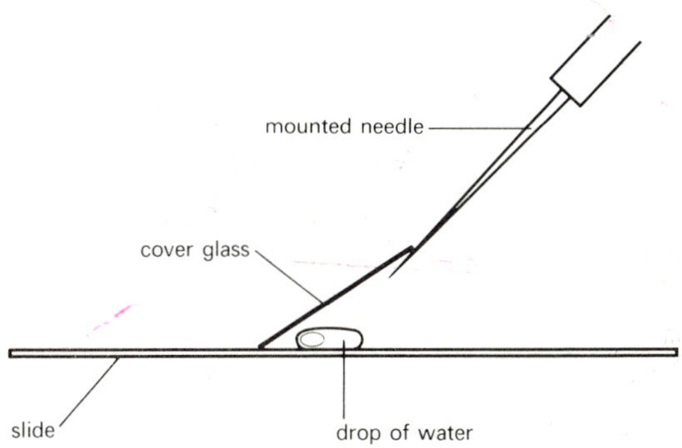

fig. 27 How to mount a specimen in water. The water should spread out to cover the area beneath the cover glass and any surplus should be absorbed by a strip of blotting paper.

very small piece of the colourless skin and mount it in a drop of water as before. Holding a clean cover glass by the rim and supporting it with a mounted needle, carefully lower the cover to touch the drop of water (fig. 27).

If there is insufficient water or if the object is too thick, air bubbles may be trapped beneath the cover glass. It is important to recognise these when examining the slide. Air bubbles are frequently trapped between fine hairs in both plant and animal specimens and can be confusing.

Once the slide has been prepared it should be examined at some length under both the low power and high power objectives and over the whole area of the slide. Investigate the effect of changing the side and angle of the sub-stage mirror and the effect of changing the size of the aperture in the diaphragm, if there is one.

From this experience it will be obvious that the image is seen best when the specimen is only one cell in thickness. Cutting stems and roots, twigs and leaves into such very thin sections is difficult indeed, and experience is needed to do it well. In many cases the greater part of any section cut by hand will be too thick for the whole of it to be clearly seen, but some small part of it may be thin enough to use. You can practise with stems of a variety of plants such as marrow and sunflower, grasses, pelargoniums and young twigs of privet and elderberry. Hold them securely in the left hand and cut across with a scalpel to give a flat cutting surface. Then cut the sections with a sharp razor blade held in a special holder or with an old fashioned 'cut throat' razor. The cutting edge of the razor should be kept moist, as should the specimen. After making a cut, any fragments of stem should be removed from the razor with a fine camel hair brush and put into a watch glass or other small dish of water. It is rarely possible to cut right across the stem nor is it necessary for much can be seen from a very small piece provided that it is sufficiently thin and is at right anlges to the length of the stem. A number of sections should be cut and placed in the water. Select the thinnest for mounting and remove them with a needle to a drop of water on a slide, cover and examine as before.

Some roots, thin stems and practically all leaves cannot be held securely enough to make cutting thin sections possible. It helps a great deal if such specimens are held in a piece of carrot or elder pith and this is cut across at the same time. Cut down the middle of a piece of carrot of a convenient size to hold between the thumb and fore-finger of the left hand, or a piece of pith about 25 mm long. The specimen, leaf or root, with the area to be sectioned facing upwards, is placed between the two halves. The carrot or elder pith holds the specimen securely and the razor cuts through both. As before, everything must be kept moist. Remove the sections of carrot or pith and the specimen to the watch glass, using the paint brush. Lift out sections of the specimen for examination and mount on a slide.

A section cut across a stem or root is called a transverse section. It gives only one aspect of its structure and sections must be cut longitudinally, both along a radius (radial sections) and at right angles to the radius (tangential sections) to give a complete picture of the arrangement of the tissues (fig. 28). These too are difficult to cut and it is sometimes easier to see the structure of the cells by a

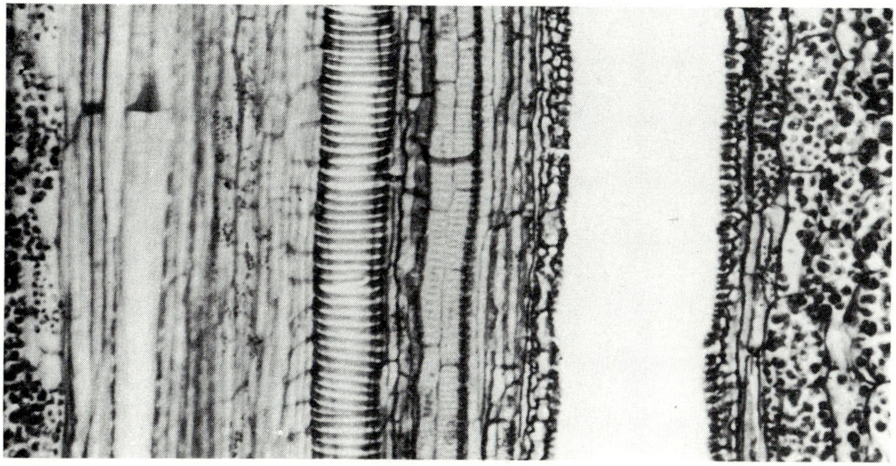

fig. 28 Longitudinal section of a stem at a magnification of approximately ×90

process called *maceration*. The sections can be more than one cell in thickness and as they are cut they are placed in a small dish of macerating fluid. This can be made by mixing 10% nitric acid with 10% chromic acid, or may be obtained from biological suppliers. The sections are left in this for approximately 24 hours, are washed in tap water, put on to a slide in water or dilute glycerine, covered and then gently tapped with the end of a mounted needle or other instrument to separate the elements present in the plant tissue. When examined through the microscope, the individual cells will be seen to lie in all directions and the variety of their size and structure can be seen.

The structure of leaves can be studied in a similar way. Put some small tough leaves such as box, laurel or privet into a beaker containing a 5% solution of caustic potash or 5% caustic soda. Boil them gently for 10-15 minutes when they will appear to be swollen and blistered. Remove the leaves, place them on a tile and cut pieces from them with a sharp scalpel. Then put these pieces into a beaker of cold water, and you will be able to observe separation of the upper and lower epidermis, leaving the mesophyll intact. These can be separately mounted on slides and examined.

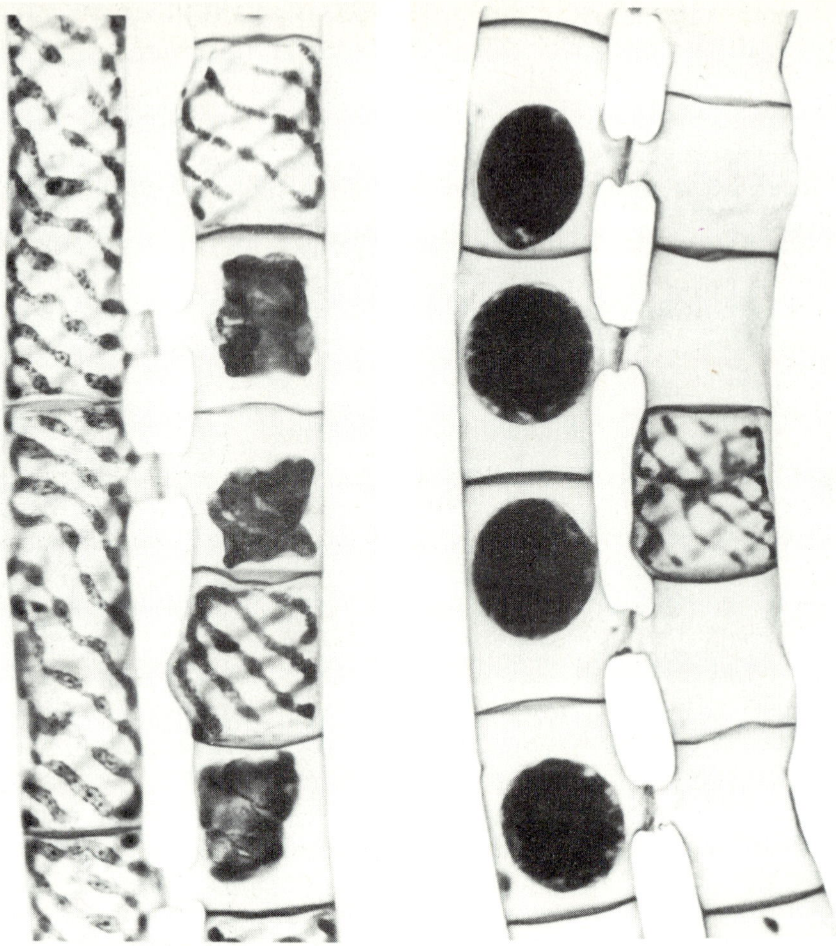

figs. 29 & 30 Spirogyra at different stages of conjugation at a magnification of approximately ×90

A never-ending source of material for those interested in the microscope is to be found in water taken from a pond or slow-moving stream or even from a deep foot print in a farmyard. A small jar of green 'weed' will provide a great variety of plants and animals for observation. A small quantity of what may look like green slime is placed in the centre of a slide and is gently teased out by a pair of mounted needles before being covered with a cover glass. The green filaments revealed will be made up of short or long cells, may be branching or unbranched and may show the presence of reproductive structures (figs. 29 and 30). Almost certainly a number of single-celled plants and animals will be visible in a state of motion. Reference to books will help to identify them.

To prevent the tiny pond animals being crushed by the cover glass, pond water should be mounted in glass slides which have a shallow depression in the centre. These are called cavity slides. Two or three drops of pond water are needed to fill the cavity before the cover glass is placed on top and the slide examined. Inevitably the speed of movement of these creatures across the field of view is too great to allow for indentification or for prolonged study to be made. Some means of slowing up their movement must be found. The addition of a little glycerine or some sticky liquid such as polycel or the proprietary 'manucol' will do this without killing them. The movement of bristles, digestive tracts and circulatory systems can then all be observed at leisure, for many of the animals are transparent.

So far, only easily obtainable living plants and animals have been considered for examination with the microscope, but the observer will still know little about the material of which the different cells are composed or the nature of the cell contents. A number of stains help to differentiate between them, but inevitably the structures being examined are destroyed by the staining process itself. A simple example of the use of stains is the reaction of starch grains to treatment with a solution of iodine in potassium iodide. Mount a thin section of a piece of potato on a slide and cover with a cover glass. Grains of starch are visible under the microscope. Dilute a few drops of iodine about 20 times with water until the solution is a pale straw colour. By means of a teat pipette, place a drop of this solution at one side of the potato preparation. It can be drawn beneath the cover glass by placing a small strip of blotting or filter paper at the opposite side. You will now be able to observe that each starch grain has taken on a bluish-black colouration. The presence of starch in any plant or animal material can always be revealed by this stain. Similarly, other materials will take up stains of different kinds: fatty globules stain reddish-yellow with a stain called Sudan III, lignified tissue which helps to give rigidity to many plant stems stains yellow with aniline hydrochloride or sulphate and nuclear tissue is stained black with haematoxylin.

Some stains are only effective when the material has been 'fixed', that is, when the tissues have been rapidly killed and preserved. A most commonly used fixative is a mixture of 99 parts of 70% ethanol and 1 part of glacial acetic acid. The material is left in this for several hours and is then washed in ethanol. It can be kept in ethanol until required for sectioning and staining. As the material is now permeated by ethanol, hand cut sections must be mounted in this medium and stained with ethanol-based stains.

Prepared microscope slides bought from biological suppliers will almost certainly have been made from material fixed, sectioned, stained and mounted by mechanical means. A microtome can be used to cut very thin sections in much greater numbers and with much greater precision than is possible by hand. Such sections are usually mounted in a medium called Canada balsam. Because this medium sets very hard when cold, the cover glass is very firm on the slide. In addition, the cover glass is sometimes ringed with a black cement to prevent any solvent from getting under it. Such mounts are called permanent preparations and are available for sale in great variety.

Smears and squashes

The examination of animal tissues such as blood, nerve tissue and nuclei requires a different technique. The blood of a frog or of man is examined by means of a blood smear. In the case of a frog, a sample of blood is obtained by inserting a teat pipette with a fine nozzle into the heart wall of a recently killed specimen and withdrawing a small quantity of blood. This is transferred to one end of a clean dry slide. To extract human blood, the skin should be carefully sterilised and a puncture made with a sterile lancet. (Details are given in the Schools Council publication, *Recommended practice for schools relating to the use of living organisms and material of living origin,* English Universities Press, 1974.) The margin of the ear lobe is suggested, and the blood allowed to drop on to a sterile slide or transferred to it by means of a sterile glass rod with rounded ends. In each case, a second slide or square-sided cover glass is used to draw the blood drop along the length of the slide so that it forms a thin smear (fig. 31). The slide is then dried by waving it in the air. It is fixed and stained by putting a few drops of Leishmann's stain or haematoxylin on the slide by means of a pipette. The stain is left for a few moments and then washed off with distilled water. The slide is dried again in air and can now be examined. It it is to be kept for any length of time, a few drops of euparal or Canada balsam must be put on the slide before covering with a cover glass.

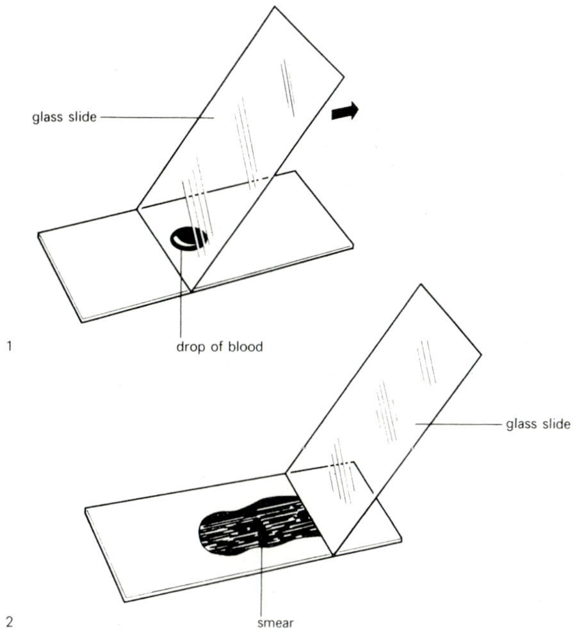

fig. 31 How to prepare a slide of a blood smear

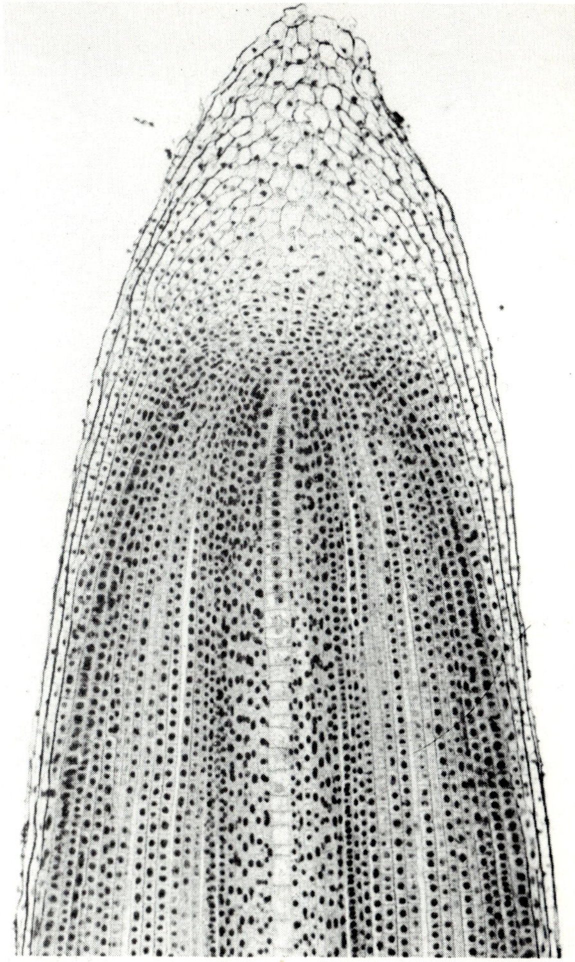

fig. 32 Longitudinal section of a root tip at a magnification of approximately ×40

Actively growing cells from the tip of a root (fig. 32) or cells from the testes of the locust can be used to show nuclei in various stages of division. Broad bean seeds are easily grown and small pieces of root are progressively cut off from the tip. They are macerated by putting into 5-10 cm^3 of N hydrochloric acid and warmed to 60°C for several minutes. They are then washed in water and each piece is put on a slide with a nuclear stain. Acetic-orcein is a suitable stain for this purpose. The root tip material is then broken up with a needle and a cover glass is placed over it. A piece of blotting paper or filter paper is wrapped round the slide and the cover glass gently pressed to squash the material below. The slide can then be sealed with rubber solution to prevent evaporation and drying up.

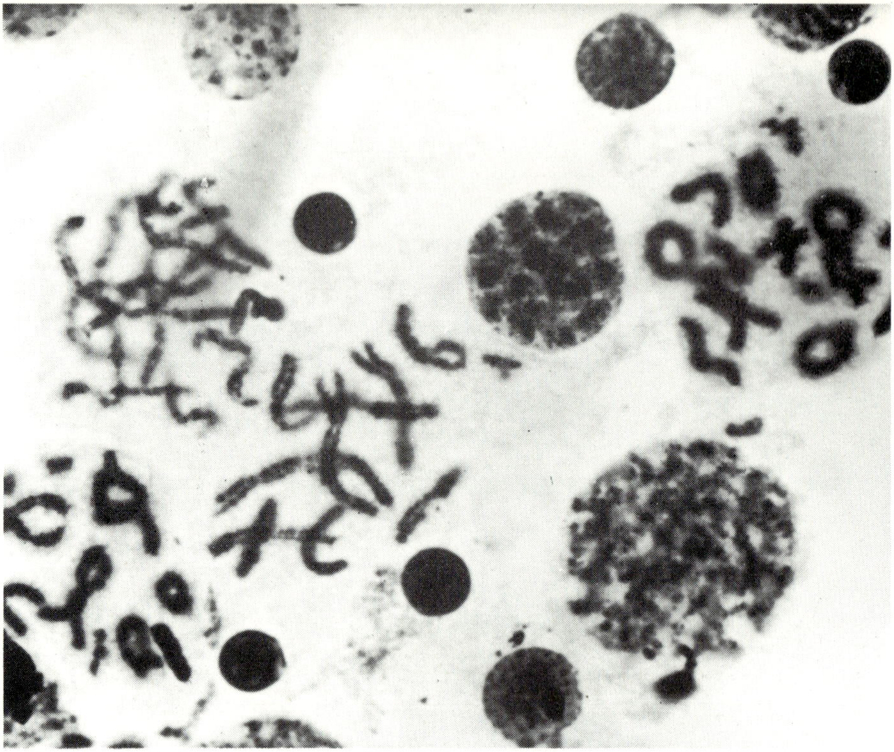

fig. 33 Locust testis 'squash' showing nuclear division

The testes of the locust can be extracted, mounted in water together with a few drops of acetic-orcein and then gently squashed in the same way (fig. 33). It is important to remember that these 'squash' preparations are only temporary ones. If the slides are to be made into permanent preparations they must be dehydrated by passing through a series of ethanols and finally mounted in Canada balsam and labelled.

Only the barest outline of the preparation of objects for microscopic examination is possible in a book of this nature. Details of the methods and techniques of fixing, staining and mounting for permanent collection can be obtained from books dealing specifically with these subjects. But whatever method is used for their preparation, it is essential that all slides being retained should be labelled with the name of the object, any stain used in its preparation and the date. Slides can safely be stored vertically in boxes or flat on trays.

5 Some useful equipment

Equipment

Once the microscope is available, the equipment needed to make good use of it can be obtained very cheaply. The most important items are slides and cover glasses, simple instruments for the preparation of specimens, and preservatives and stains.

Collecting

Plant and animal material provides an infinite variety of objects for study with both lens and microscope. Flowers and leaves should be kept fresh in a plastic bag or similar air-tight container.

To collect ditch, pond or stream water, a few jam jars, especially those with screw top lids to prevent spillage, are sufficient. A single jar full of water from a sluggish stream or stagnant pond and containing weed of any kind will provide more than enough for study over a considerable period. A special nylon or fine silk net with a metal tube attached, called a plankton net, is needed to skim the surface of a large area of open water.

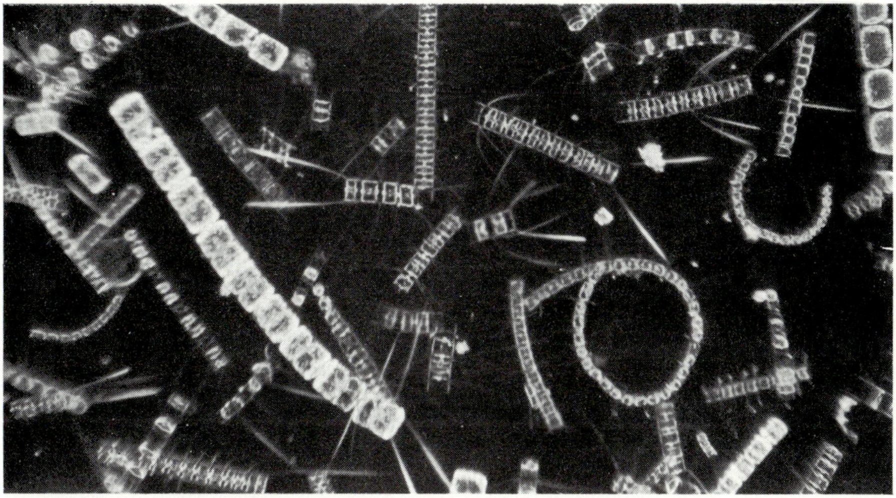

fig. 34 Plant plankton at a magnification of approximately ×45. They were collected with a plankton net which was dragged behind a boat to cover a reasonable area at a time. The material shown in fig. 16 was also collected in this way

Instruments

These will include forceps, fine scissors and scalpels, small paint brushes and mounted needles. You can buy scalpels of several sizes, some with wooden handles, others entirely of steel. A useful model consists of a steel handle and removable blades in a variety of sizes and shaped cutting edges. Mounted needles may be bent or straight and are especially useful for teasing out material which has already been put on the slide. A razor is essential for making plant sections.

Slides and cover glasses

Microscope slides are of standard size, 75 mm ×25 mm and they are sold in boxes of 100 slides. For general use they are between 0·8 and 1·8 mm thick. In addition to these, cavity slides and others with a shallow glass ring fused to the centre can be bought. Both make it easier to examine small pond creatures without any danger of squashing them beneath the cover glass.

Cover glasses may be round or square in shape and these too can be bought in a number of different thicknesses ranging from 0·085 mm to 0·35 mm. Because they are so easily broken great care has to be taken in handling them. They are sold in boxes of 100.

A recent development is an aerosol spray which can replace cover glasses on some microscope preparations. It is sprayed on to the specimen and dries in less than ten minutes to a non-sticky, glass-like finish. It can be washed off again with xylol.

Staining and mounting

Teat pipettes are most useful in ensuring that water, glycerine, ethanol and stain are applied to the slide drop by drop; you will need several of these.

Fixatives and stains bought in bulk should be transferred to dropping bottles, each holding 30 cm^3 and fitted with a teat pipette. If permanent preparations are being made, ethanol in concentrations of 70% , 95% and 100% (absolute) will be needed. Xylol will be required before the final mounting and also for the cleaning of slides and lenses. Canada balsam is sold in special bottles with an inverted cone as an air-excluding lid. As only a very small quantity of this sticky liquid is needed at one time, a solid glass rod is supplied with the bottle and is used to transfer a drop of it to the slide.

To avoid spilling liquids on to surfaces which might be damaged by them, work with stains, ethanol, xylol, etc, is best done over a ceramic tile or a sheet of glass which can be wiped clean. Watch glasses or other small dishes are useful for holding thin sections and for many other purposes. A soft cloth is essential for keeping everything clean and free from dust.

Labels for identification and boxes for storage are required for permanent collections.

6 Other uses and accessories

The microscope in physics

It is difficult to understand the nature of pressure without actually observing the behaviour of particles in motion. This behaviour is most effectively demonstrated by using a microscope to show the movement of smoke particles. The distance between the microscope stage and the objective lens must be sufficient to take the glass 'cell' in which the smoke is confined, and not all microscopes are suitable. The lower power objective is adequate and a ×10 eyepiece should be used. Smoke is obtained by lighting one end of a drinking straw. The smoke travels down through the centre of the straw and can easily be directed into the smoke cell. When it is full, a cover glass is used to close the opening and the slide is placed in the centre of the stage. Light is provided by a festoon bulb connected to a 12 volt supply and is directed through a cylindrical glass rod, which acts as a lens, and then passes horizontally through the smoke cell. Focus the microscope first on the cover glass and then carefully down until the movement of the smoke particles can be seen as tiny illuminated specks.

The microscope in geology

The formation of crystals is a fascinating sight which can be observed through the microscope. Saturated solutions of substances such as barium choloride, potassium dichromate and copper sulphate can be used. Put a few drops of the solution on a slide and you can observe the crystallisation process as the water evaporates. The recognition and identification of some of these crystalline substances is very important in the study of minerals. In the case of rocks which are composed of several minerals, the microscope is most useful in determining their nature and relative frequency.

Some detail is revealed by light transmitted through a thin section of rock mounted on a slide in the same way as the biological specimens, but much more is visible if polarised light is used. This is a beam of light in one plane only, the rest of the light having been cut out in some way. The earliest way of doing this was to pass the light through a prism of Iceland spar. This substance possesses the unique quality of allowing light to pass through it in one direction only. The beam of light is said to be *plane polarised* and the prism is called a *polariser*. If this beam is then passed through a second similar prism placed at right angles to the first, the light is cut out entirely. This second prism is called the *analyser*. If a thin section of rock is placed between the polariser and the analyser, the direction of the plane polarised light passing through it is changed. The extent of this deflection can be used in determining the nature of the rock. If this deflected beam is then passed through the second prism, the analyser, the refractive properties of the mineral are displayed and the constituents of the rock are revealed by diagnostic colours.

In many microscopes a polariser can be fitted below the stage and the light directed through it from the sub-stage mirror. The analyser is usually fitted into the microscope tube and can be pushed in or out with a slight movement of a finger.

The polarising version of the Open University microscope makes use of small pieces of material called polaroid which behave in the same way as the prisms of Iceland spar.

To obtain the best results from a polarising microscope, the objects must be prepared for study by carefully cutting and grinding small pieces of rock. Each piece is ground and polished to give a flat surface and this is then fixed to a piece of glass by means of Canada balsam. The opposite side is then ground flat until the section is as this as possible. The cementing balsam is melted and the section is removed, cleared by immersion in xylol and then remounted in Canada balsam on a clean slide and covered with a cover glass.

Micro-projectors

The micro-projector or projection microscope enables many people at one time to see an image of the enlarged slide. There are a number of models available, some projecting the image on to a wall screen, others on to a small ground-glass screen in a detachable head which can be fitted on to a standard instrument. The clarity of the image depends upon both the illumination of the specimen and the intensity of the light in which the image is projected.

Photomicrography

The camera is a most useful instrument for recording what is seen through the microscope. Many models of camera can be used for this purpose and it is possible to buy adapters which fix the camera in position closely over the eyepiece. Several of the illustrations in this book were taken in this way (figs. 11, 12, 16, 33).

Book list

General reading

The Revealing Lens, Brian J Ford, Harrap 1973
Through the Microscope, M Dampier Anderson, Aldus Books 1965
Introduction to Microscopy, George W White, Butterworth 1966
Modern Microscopy, C F Culling, Butterworth 1974

Techniques of slide-making

Peacock's Elementary Microtechniques, revised by Savile Bradbury, Edward Arnold 1973
A Guidebook to Microscopical Methods, Grimstone and Skaer, Cambridge University Press 1972

Photomicrographs and drawings

The Protozoa, Vickerman and Cox, John Murray 1967
Atlas of Histology, Freeman and Bracegirdle, Heinemann 1966
Atlas of Embryology, Freeman and Bracegirdle, Heinemann 1967
Atlas of Invertebrate Structure, Freeman and Bracegirdle, Heinemann 1971
Atlas of Plant Structure vol 1, Bracegirdle and Miles, Heinemann 1971
Atlas of Plant Structure vol 2, Bracegirdle, Heinemann 1973
Photomicrographs of the Flowering Plant, Shaw, Lazell and Foster, Longmans 1965
Photomicrographs of Non-Flowering Plants, Shaw, Lazell and Foster, Longmans 1968

Suppliers

General—For microscopes, biological supplies, mounting materials and reagents.

Griffin & George, 285 Ealing Road, Wembley HA0 1HJ
Philip Harris Biological Ltd., Oldmixon, Weston-Super-Mare, Avon, BS24 9BJ

Optical Instruments, especially microscopes.

Bausch & Lomb Optical Co. Ltd., Lenten House, Lenten Street, Alton, Hants GU34 1JB
Gillett & Sibert Ltd., 417 Battersea Road, London S.W.11.
Micro Instruments (Oxford) Ltd., 7 Little Clarendon Street, Oxford, OX1 2HP. *Meopta* models.
Opax, 142 Silverdale Road, Tunbridge Wells, Kent TN4 9HU.
Charles Perry & Co., 192 Ilford Lane, Ilford, Essex.
W.R. Prior & Co. Ltd., London Road, Bishop's Stortford, Herts.
Pyser Ltd., Fircroft Way, Edenbridge, Kent TN8 6HA. *Swift* models.
Scientific Optics Ltd., Ponswood Industrial Estate, Hastings, Sussex TN34 1YA. *Open University* models.

Index